'ARK

ɔute

Amanda Le Boutillier
Michael Tooley
Robin Stannard
Nick Baxter
Lizzie Noble

Published by The Society for the Arts and Crafts Movement in Surrey 2012

Printed by Synergie Group, Lymington, Hampshire SO41 8LZ 0845 520 8000

Additional Editorial Work: Denise Todd and Hilary Underwood

ISBN 978-0-9537615-6-2

The Society for the Arts and Crafts Movement in Surrey exists to celebrate and foster interest in all forms of art, architecture and design of 'the Arts and Crafts Era' - the period from 1860 to 1930. The Society is concerned with the conservation and awareness of works associated with the period and is gathering together information such as publications, photographs and drawings relating to this extraordinary period of creativity in Surrey.

The Society was formed in 1996 to act as a forum for those who share this interest and enthusiasm. Members enjoy a programme of lectures, usually held at the Watts Gallery at Compton, study days and visits to places of Arts and Crafts interest and receive the Society's newsletters, issued twice a year. New members are always welcome and there are plenty of opportunities to assist with the Society's projects for those who wish to do so.

www.artsandcraftsmovementinsurrey.org.uk

Front cover: Phillips Memorial Cloister. (Photograph: Sarah Sullivan)
Back cover: 1910, Jack Phillips, photographed by Jenny Stedman. The Godalming Free Press

CONTENTS

The Phillips Memorial
Cloister (Drawing
Sarah Sullivan)

INTRODUCTION

Jack Phillips was a young man who, on the night of 14th April 1912, just four days after his 25th birthday, did something extraordinary. His calm devotion to duty, his courage and self-sacrifice, were seen as remarkable and worthy of recognition at the time, and still inspire people today. This absorbing book tells the story of Jack Phillips, the memorial to him in Godalming and his enduring legacy.

Amanda Le Boutillier and John Young explain how the boy born above the draper's shop in Farncombe, came to travel the world during an exciting career which culminated in the prestigious posting as Chief Telegraphist on the *Titanic*.

Jack Phillips is someone we can identify with, and perhaps part of the fascination of his story is that we ask ourselves what we would have done that night, if we had been in the Marconi 'shack' on the *Titanic*, responsible for the only means of communication which might bring rescue to the 2,224 people on board.

Because Harold Bride, the Assistant Telegraphist, survived, we know what Jack did; he stayed at his post, sending the increasingly desperate wireless messages with a steady hand. After Captain Smith released them both, Jack Phillips sent Harold away, but made no attempt to save himself. His last message was received at 2.17am and the ship went down three minutes later. 1,519 People died in the disaster, but Jack's radio messages brought the *Carpathia* to rescue the 705 survivors.

The scale of the Phillips Memorial Cloister and Grounds, between Godalming Parish Church and the River Wey, is an indication of the depth of feeling aroused by the heroic and tragic death of this young man. It is also a testimony to the vision, skill and determination of the Committee for the Design of the Phillips Memorial. In Sarah Sullivan's essay you can meet the remarkable women who formed this committee and discover how their involvement in key social and artistic movements of the times influenced their work. Michael Tooley examines the central role played by the garden designer Gertrude Jekyll and Robin Stannard the contribution of the architect, Hugh Thackeray Turner. These essays, together with the historical perspective given by Russell Morris's essay, allow the Memorial to be assessed in the context of the Arts and Crafts Movement.

The Phillips Memorial was opened with a service of remembrance on 15th April 1914. Four months later, Britain entered the First World War. The whole nation would have to find ways to commemorate the courage and loss of a generation of young men. In Godalming, the War Memorial looks across the Phillips Memorial Ground to the Phillips Memorial Cloister, integrated by planting designed by Gertrude Jekyll.

The War Memorial was the first, but by no means the only, addition to the Phillips Memorial. In their essay, Nick Baxter and Lizzie Noble examine how the Cloister and the Park have matured over the past century and explain how Waverley Borough Council, with the support of the National Lottery, continues to honour the original vision - that Jack's memorial should enrich the life of the community as well as enhance the appearance of the town.

Alison Pattison
Curator, Godalming Museum

Iceberg memorial stone to Jack Phillips. Nightingale Cemetary, Farncombe, Godalming (Photograph: Sarah Sullivan)

THE ESSENCE OF AN ARTS AND CRAFTS BUILDING

Russell Morris

An overview of the origins of the Arts and Crafts Movement, and an attempt to define its architecture, is a useful starting point in the story of the Phillips Memorial Cloister and its setting in the Phillips Memorial Park.

RMS *Titanic* was, in aspiration, a state of the art ship. Its outward appearance, informed by its function, anticipated the moderne. The interior, at least for first class passengers, was that of an opulent historic country house. This contradiction, if unconcerned with the niceties of aesthetic purity, was undoubtedly good commercial thinking. *Titanic* looked exciting, fast and unsinkable - but offered traditional palatial luxury within.

At home, though, a more contemplative approach to design had firmly captured the thinking classes: the Arts and Crafts Movement. Since proponents included Gertrude Jekyll and Hugh Thackeray Turner, it was inevitable that its principles guided their design of the Phillips Memorial. But what is the essence of the style?

The Arts and Crafts Movement emerged around 1860, but to understand its origins we need to start just a little earlier in history. Britain prospered with the Industrial Revolution.

FIG. 1
THE GREAT EXHIBITION
(TALLIS'S HISTORY AND DESCRIPTION OF THE CRYSTAL PALACE AND THE EXHIBITION OF THE WORLD'S INDUSTRY IN 1851, LONDON JOHN TALLIS AND CO. 1851, VOL 1, OPPOSITE P. 44)

Hugely increased scope for mass production enabled manufacturers to serve vast markets at home, in the Empire and beyond. Even if this new prosperity was not fairly shared, it was a source of a national pride that was eventually to inspire the Great Exhibition of 1851, a triumphant celebration of Britain's achievements. (Fig. 1) Modern manufacturing offered elaborately decorated goods at a price that made them available, if not to the poor, at least to the growing middle class. If for some this was a matter of pride, for others it was a cause for reaction. In the decades that followed, artists and intellectuals questioned the vulgarity of facile ornamentation and the loss of a more worthy pride in craftsmanship. Machines might have their place in reducing arduous labour, but campaigners believed that they should remain at the command of craftspeople, each of whom should retain a responsibility for the finished product rather than solely one small stage of manufacture.

FIG. 2
WILLIAM MORRIS, DESIGNER, WRITER, TYPOGRAPHER AND SOCIALIST, 1884. PHOTOGRAPH BY FREDERICK HOLLYER (WATTS GALLERY ARCHIVE)

Concern for the dignity of labour, expressed by influential thinkers, such as John Ruskin and William Morris, (Fig.2) generated the Arts and Crafts style and also a flourishing social and artistic movement that survived until the Great War and beyond. A wide range of products became available, including: wallpapers, fabrics, furniture, silverware, pewter and glass, often of mediaeval, natural or folk inspiration. (Fig.3) Liberty & Co established itself in 1875 as the flagship retailer of goods of refreshing simplicity and quality, but costly and well beyond the reach of ordinary people. The Movement quickly spread worldwide finding a particular hold in America.

Perhaps the architectural Arts and Crafts springboard was Red House (1859) designed for William Morris by Philip Webb. (Fig.4) This, with its simple materials and reliance on honest craftsmanship, exemplified Morris's philosophy. There is more than a nod to mediaeval form, but it is nevertheless a timeless building unconstrained by rigorous adherence to any historic style or fashion. It is simply the house that Morris wanted to suit his own lifestyle and ideals rather than those of conventional society.

For the next half century or so architects embraced and developed this approach with enthusiasm, including (to mention some of the most important): Edward Prior, Mervyn Macartney, Hugh Thackeray Turner, CFA

FIG. 3
COBRA POT, DESIGNED BY MARY WATTS (WATTS GALLERY ARCHIVE)

Fig. 4
Red House (1859)
designed by Philip Webb
(Drawing John Fairbanks)

Voysey, William Lethaby, Gerald Horsley, MH Baillie Scott and of course Sir Edwin Lutyens. They were encouraged by a good harvest of newly-rich clients anxious to demonstrate their own artistic credentials. The heathlands of Surrey, if poor for agriculture, were easily accessible from London so proved particularly fertile ground for Arts and Crafts homes for the affluent. Here, particularly with the early works of Lutyens, (Fig.5) (Fig.6) was the epicentre of the Movement which reached its zenith between circa 1885 and the outbreak of war in 1914.

The construction of an Arts and Crafts building was not, of course, a cheap exercise. But there were crumbs for the poor as landowners built prestigious estate cottages and lodges for their workers. Social reformers, too, adopted the style in the design of Garden Cities bringing benefits to a broader suburban population.

Fig. 5
Tigbourne Court,
Witley (1899) designed
by Edwin Lutyens
(photograph: John
Young collection)

What are the stylistic characteristics of an Arts and Crafts building? Stemming from the principle of respect for craftsmanship there should be a clear exposition of sound, honest construction. That means good materials, well-considered detailing, usually following traditional local vernacular, and a response to the physical setting. There should be originality, but even the best architects sometimes raided their old sketchbooks for picturesque features that they could weave into their designs. They did, though, reinvent them and combine them in new ways rather than slavishly pin them to their buildings as mere decoration. Again, like Red House, this so often gives a timeless quality to a building – making it gloriously independent of any style or period.

Fig. 6
Goddards, Abinger Common (1900) designed by Edwin Lutyens (photograph: Sarah Sullivan)

Buildings of most other periods of British architecture can be identified by distinct stylistic features, be it the shape of an arch, the design of a window or the display of a fashionable decorative form. Victorian architects often revived these historic styles but their work is usually identifiable by a rigorously academic, but sometimes sterile, application of features or by mere symbolic pastiche. Arts and Crafts buildings though, certainly the best of them, are typified by highly creative and resourceful use of traditional ideas. They display originality that takes them beyond the mere eclectic. Features borrowed from traditional buildings are combined in novel compositions to highly picturesque but still practical effect. Originality is the key and it is unsurprising that some architects, very capable exponents of Arts and Crafts design, worked happily in other styles too. Oliver Hill built numerous Modernist houses; Goodhart-Rendel (very much at home building cottages on his own Surrey estate) designed the remarkable Art Deco St Olaf House at Hay's Wharf.

The Arts and Crafts Movement grew from a strong purist philosophy but the best examples of its architecture were eventually raided for ideas by speculative builders seeking a garb for their own less original creations, which were not so well grounded in an understanding of the original sources. The Arts and Crafts ideal was reduced to a fashionable style suited to the expanding, polite suburbs. Occasionally, though, even these buildings display some naïve charm.

The real gems of the genuine Arts and Crafts Movement are distinguishable by their coherent composition and quality of build. It is not easy to define what constitutes an Arts and Crafts building, it is more elusive perhaps than any other style – but its essence is immediately recognisable.

It is not possible in just a few paragraphs fully to describe the Arts and Craft Movement. For proper coverage of the variety and complexity of the subject, Peter Davey's *Arts and Crafts Architecture* is an excellent and lavishly illustrated book. For a briefer account, with a Surrey focus, *Nature and Tradition* is the ideal source. [See the bibliography.]

JOHN GEORGE (JACK) PHILLIPS (1887-1912)

John Young and Amanda Le Boutillier

One of the most enduring legends of heroism on the *Titanic* started in a flat above the Farncombe branch of Gammon Bros, the Godalming drapery business. (Fig.1) It was here, on 11th April 1887, that Anne, wife of the shop manager George Phillips, gave birth to their only surviving son, John George, known as Jack. He was a much longed-for heir, arriving after the loss of two other children, when his parents were in their mid-forties and his sisters, Ethel and Elsie, were already in their teens.

Jack was christened at St John the Evangelist Church in Farncombe on 29th May 1887. (Fig.2) His family were regular members of the congregation and Jack later sang in the Church choir. His early education was at the Church of England School, next door to St John the Evangelist, where he was taught alongside a number of his contemporaries in the choir. (Fig.3) Several of them became lifelong friends who remembered him with affection. The remainder of his schooling took place at Godalming Grammar, at the time a small private school housed in the meeting hall next to the Red Lion Hotel, a room which is now the Red Lion's public bar. (Fig.4) The Headmaster of the school, Charles Elworthy, later spoke of him as 'a lad of pleasant disposition who was well liked by his school fellows'.

Jack struggled with his studies and, as a result, his father arranged additional lessons for him with a local private tutor, Mr Mandeville. Having left school, Jack became a telegraph boy at Godalming Post Office, (Fig.5) delivering telegrams around Godalming and Farncombe on his bicycle. (Fig.6) At the same time he studied for the Civil Service Post Master General's examination and, having successfully passed and reached the required

Fig. 1
circa 1910: Gammon Bros, Farncombe (John Young collection)

speed for sending and receiving Morse code, he became a trainee telegraphist.

FIG. 2
ST JOHN THE EVANGELIST CHURCH, FARNCOMBE (PHOTOGRAPH: SARAH SULLIVAN)

After four years, Jack left the Post Office and travelled to Liverpool to study at the Marconi Company Training School with the intention of going to sea. Having discovered his talent, Jack became completely absorbed in the exciting world of wireless telegraphy in much the same way as today's computer 'geeks' love I.T. He was remembered by his instructor, Mr Blinkhorn, as 'a pleasant boy, well spoken, good tempered and friendly'.

By August 1906, Jack had completed his training and went to sea for the first time as Marconi Operator of the White Star Liner, *Teutonic*. For the next couple of years he happily criss-crossed the Atlantic between Liverpool, New York, Montreal, Portland and Quebec. He served on ships from the White Star, American and Cunard Steamship Company, including *New York, Lusitania, Mauritania* and *Oceanic*. The latter was his favourite ship and a tour of duty that he always particularly enjoyed. He had become, according to a Marconi official interviewed later by the Daily Express 'One of our most trusted and efficient servants'. (Fig.7)

Throughout his travels, Jack remained close to his family, visited whenever he could and wrote regular letters about his adventures. Knowing that his sisters liked to collect the pictures, he sent a weekly postcard, usually to Elsie who was collecting them in a green postcard album, but also some to Ethel. Ethel, who suffered ill health throughout her life, remained living at home with her parents, helping in the shop and housekeeping. Elsie had left home and was living in Ripley, working as a teacher at the Ryde House School. Her postcard album survives and although the cards were separated following an auction in 1997, several examples will be on display at Godalming museum during the 2012 commemorations.

FIG. 3
ST JOHN THE EVANGELIST CHURCH SCHOOL, FARNCOMBE (PHOTOGRAPH: SARAH SULLIVAN)

Between 1908 and 1911, Jack was assigned to Marconi's transatlantic transmitting station at Clifden on the West Coast of Ireland. He did not enjoy the posting and was eager to get back to sea. The station was very isolated, the weather damp and, as they lived on site, there was little to occupy the single operators, when they were off duty. But during this period, Jack developed a long distance friendship with an operator at the Glace Bay Station in Nova Scotia, Walter Gray, who was originally from the Shetland Islands.

Gray and Phillips met up when Gray came back to England on leave and Jack was on the landing stage at Liverpool docks to wave his friend off when he returned to Canada. Having been promoted to Officer-in-Charge of the Cape Race Station in Newfoundland, Gray would be forced to stand by helplessly on a freezing night in April while his old friend broadcast his desperate pleas for assistance from his most glamorous, but final posting.

Fig. 4
Red Lion, Godalming
(former High School)
(Photograph: Sarah Sullivan)

In the summer of 1911, Richard O'Driscoll, Senior Operator at Clifden, recommended Jack Phillips for a posting aboard the RMS *Adriatic* and Jack returned to sea. While aboard, he was photographed by Francis Browne, a trainee priest who he later met again on *Titanic*. The picture features Jack having a quiet cigarette with a second Marconi Operator. This man is frequently identified as Harold Sydney Bride, *Titanic's* Junior Operator. But Bride was on the Brazilian run at the time and, as Marconi's records from the time do not survive, the *Adriatic* Operator's identity currently remains a mystery.

Jack met Harold Bride in early 1912 in Belfast when they joined the new White Star Liner, RMS *Titanic*. The pair were probably the envy of their fellow Marconi Operators as they embarked on the fine tuning of the new equipment, one of the most powerful shipboard systems currently in service. On 2nd April the ship travelled to Southampton and while *Titanic* was prepared for her maiden voyage, Jack took a trip to Cowes before she left for Cherbourg on 10th April. The following day, Jack's 25th birthday, the ship paused briefly in Queenstown (now Cobh), Ireland to collect her final group of passengers. Here, Francis Browne, Jack's acquaintance from the *Adriatic*, left *Titanic* taking with him a unique collection of photographs of life aboard one of history's most famous ships.

Fig. 5
HSBC bank
(former Post Office)
(Photograph: Sarah Sullivan)

For a record of what happened in *Titanic's* Marconi shack during the ill-fated voyage, we must rely on the evidence of Harold Bride, a 22 year old with only nine months experience as a telegraphist at the time of the disaster. When, with Guglielmo Marconi's blessing, he gave an interview to Jim Speers of the *New York Times* aboard *Carpathia*, he was suffering from exhaustion, shock and badly frostbitten feet. Speer's subsequent story seems a remarkably lucid account from such a traumatised young man, but Bride had clearly developed a tremendous amount of respect and affection for Jack Phillips, a man whom he had only known for a few weeks but who had left a lasting impression.

FIG. 6
1906: GODALMING POST OFFICE WORKERS. JACK PHILLIPS IS STANDING, SECOND FROM THE RIGHT (JOHN YOUNG COLLECTION)

Bride explained the breakdown of the *Titanic's* wireless system on Saturday 14th April, the seven exhausting hours that Jack spent tracking down and repairing the burnt out transformer secondary and the considerable backlog of commercial traffic which Jack insisted on tackling himself while sending Bride to bed. Mention of the backlog of commercial traffic brings the question of the ice warnings. Phillips's handling of these has generated controversy. Of the six messages received by *Titanic*, three reached the bridge and were recorded. Of the three others, messages from *Amerika* and *Mesaba* were received by Phillips but not acknowledged, suggesting that they had not been delivered to the bridge. None of the surviving officers admitted seeing them. It has been suggested that Jack gave priority to profitable passenger traffic, but would he really have neglected ice warnings on purpose? It seems more likely that he had every intention of delivering them given a suitable break, that with only a basic knowledge of navigation he did not truly appreciate their importance and was tragically overtaken by events.

A transmission from Cyril Evans of the Leyland Liner *Californian* that Jack cut off with his angry 'Keep Out, I'm working Cape Race, you are jamming me' is seen as damning evidence by some that Jack had singlehandedly caused the *Titanic* disaster. This is grossly unfair and untrue. Evans himself did not find Jack's behaviour unreasonable. He admitted his transmission must have been very loud because he could hear *Titanic* clearly and, interestingly, he admitted that he had not used a Masters Service Gram prefix to warn Jack of its importance. It is unlikely that Jack heard the whole message as he shied away from his headphones to protect his ears from Evan's blast. Even if he did, he probably thought it was

FIG. 7
JACK PHILLIPS, CHIEF TELEGRAPHIC OFFICER ON RMS *TITANIC* WHEN IT SANK IN 1912 (1910 PHOTOGRAPH BY JENNY STEDMAN. THE GODALMING FREE PRESS)

Fig. 8
The *Titanic's* collision with the iceberg (Drawing Sarah Sullivan)

Fig. 9
Jack Phillips in the *Titanic's* radio operator's cabin. Based on historical and archaeological research (Drawing Sarah Sullivan)

just chatter and not a message for the Captain. Cyril Evans was rather notorious for jamming and unnecessary chatter on the airways and on the very same voyage had been warned about his conduct by one of Marconi's travelling Inspectors.

After spurning Evans's message and clearing the backlog, Jack briefly caught up with Walter Gray at Cape Race. When the collision occurred at 11.40 pm, he felt only a slight bump which he interpreted as a propeller being thrown. (Fig.8) He had visions of an immediate return to Belfast for repairs and was telling Bride, who had just joined him from their sleeping quarters, when Captain Smith arrived and ordered the distress signal to be transmitted. Jack gave up all thoughts of going to bed and returned to the wireless set. As the situation became increasingly grave, he left it only briefly to look outside, returning with the comment 'Things

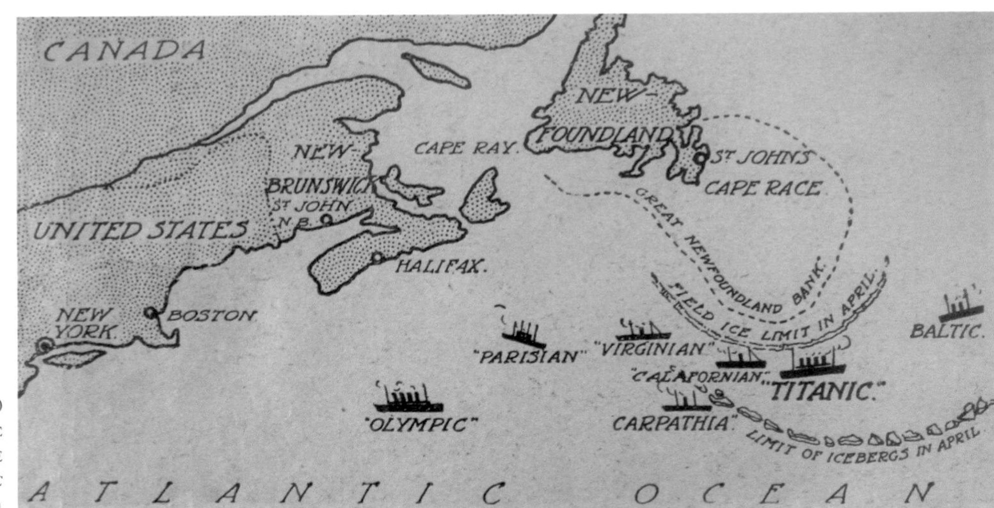

Fig. 10
Map of the Atlantic Ocean showing the position of the *Titanic* (Sarah Sullivan)

look very queer'. From then on he stayed at the wireless key calmly giving information to ships responding to his distress calls at half his normal sending speed to ensure they got the correct information. The New York Times reported that he was 'Absolutely cool and clear headed, his sending throughout being steady and perfectly formed'. (Fig.9)

Just after 2.00 am, Captain Smith released Phillips and Bride but Jack refused to leave his post. Knowing the power would not last much longer, he was determined to stay as long as it held out and try to keep in touch with the outside world. It must have been a comfort to him that the Cunarder *Carpathia* was just 58 miles away and coming with all possible speed. He tested the spark with the message 'VV CQD' which was heard by the *Virginian,* but before he could complete the message the set went dead. (Fig.10)

What happened next and how Jack met his end is still the subject of speculation. Second Officer Charles Lightoller claimed to have had a conversation with Phillips on board Collapsible lifeboat 'B'. He said he saw Jack pass away but that he had insisted the body was transferred onto *Carpathia*. However, the body retrieved from 'B' was later identified as that of Russian jeweller, Abraham Harmer. Colonel Archibald Gracie was adamant that the man talking to Lightoller was not Phillips, but as a First Class passenger and not a member of the crew, how would he have known what either of the Marconi men looked like? Harold Bride escaped on to the roof of *Titanic's* wheel house and made it on board the upturned 'Collapsible B'. Bride claimed that he had not seen Phillips alive after he left *Titanic*. Surely, on a lifeboat so terrifyingly small, if Phillips had been there, the two colleagues would have discovered each other and spoken during that period of several hours. On reflection, probably Lightoller, a 'company man' through and through

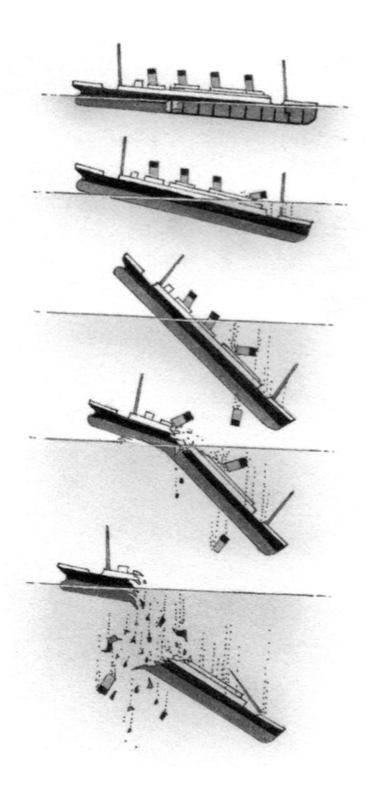

Fig. 11
Five stages of *Titanic* breaking up and sinking
(Drawing: Getty Images)

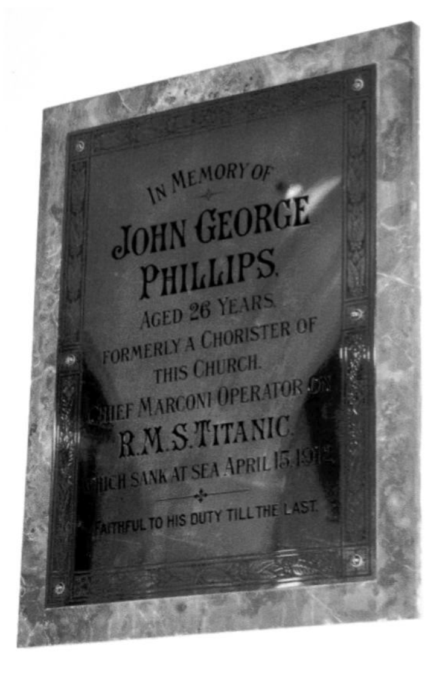

Fig. 12
Brass memorial to Jack Phillips at Farncombe Church
(Photograph: Sarah Sullivan)

wanting to defend the reputation of the White Star Line, had been mistaken as to whether he had spoken with Bride or Phillips.

The details of Jack Phillips's last moments and final resting place will never be established. (Fig.12) Like so many of the unfortunate souls aboard *Titanic*, his only grave was the icy water of the North Atlantic. But in his home town, a superb Memorial was created that ensured that his sacrifice would never be forgotten.

JACK PHILLIPS FAMILY TREE

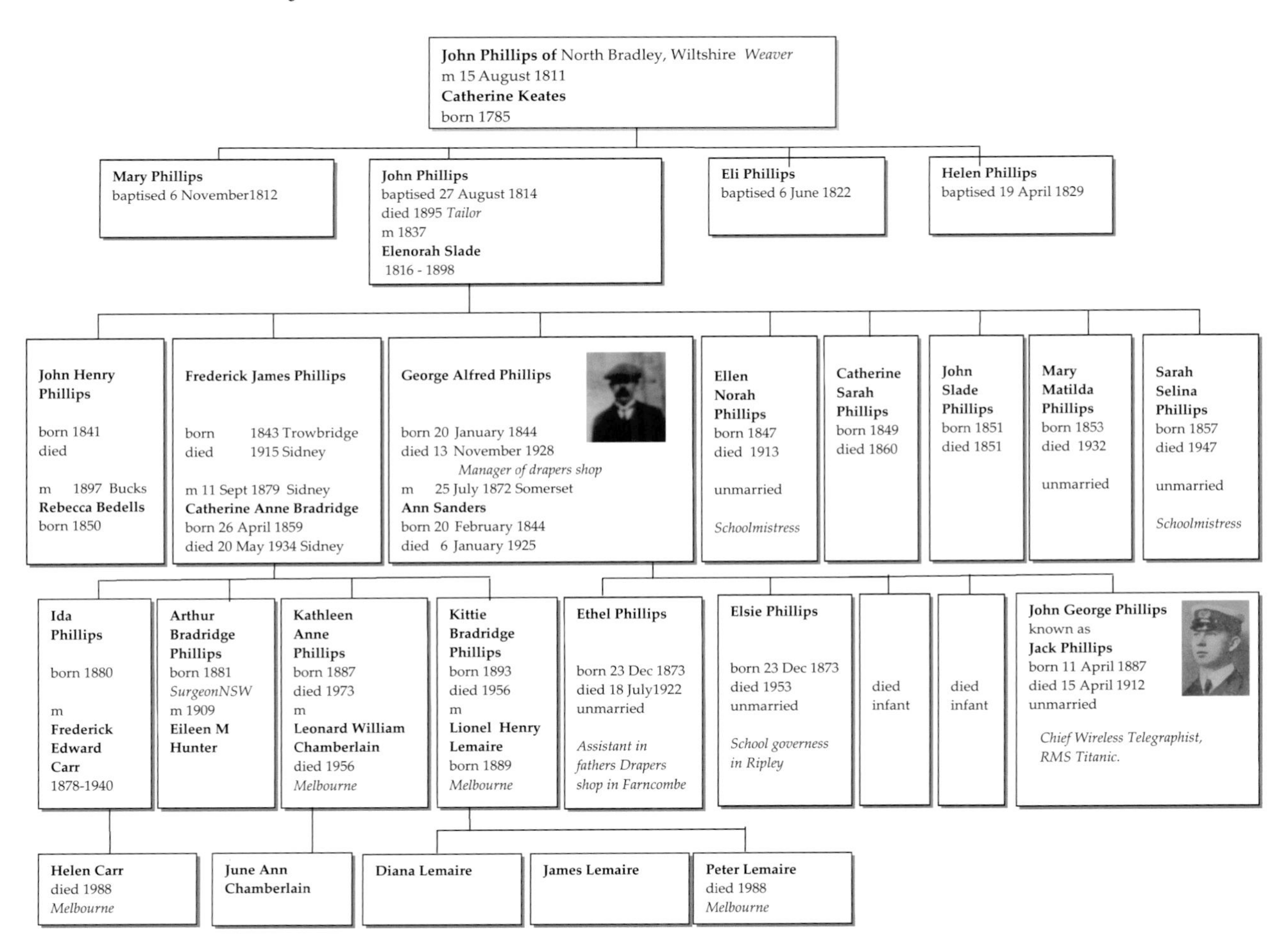

THE FIVE WOMEN INVOLVED WITH COMMISSIONING THE PHILLIPS MEMORIAL

Sarah Sullivan

The news of the *Titanic* disaster caused public anguish on both sides of the Atlantic with an outpouring of grief on an international scale. Jack Phillips was portrayed in the press as one of the heroes; his perseverance in continuing to transmit calls for help in Morse Code, had captured the public's imagination. It was only during the Inquiry held in America and Britain that a negative press report evolved, which has continued to provide controversy to the present day. The people of Godalming knew these accusations to be false and they rallied to the cause of providing a fitting tribute to their young hero.

Fig. 1
Photograph of Mary Watts and assistants creating a plaster panel (Watts Gallery archive)

The loss of the *Titanic* provoked a great deal of flamboyant writing at the time. The fact that the largest, finest and safest vessel afloat, the last word in luxurious shipbuilding, the unsinkable ocean palace was engulfed in the dark icy waters on her maiden voyage with a death roll of 1523 drowned persons speaks volumes in itself and needs no elaboration. Despite all the modern advances in shipbuilding, Nature had shown mankind her indomitable power.

By the end of April 1912, the Godalming Town Council had opened a fund to provide some form of recognition of Jack's deed to help save 705 lives at the sacrifice of his own. Donations and letters from England, America and Canada were highlighting the need for something to be constructed as a fitting tribute. Reports in the local press stated the amount raised each week with a list of subscribers and exactly how much each had contributed. Meetings and public gatherings in the Municipal buildings brought together a town upset by the flood of accusations in the press, referred to as a 'welter of hysterical unwisdom'.

There were five women involved with the design of the Phillips Memorial. They were well acquainted with each other and wanted to provide something more than just the standard unimaginative response. Initially four women formed the Godalming Town Council Committee for the design of the Phillips Memorial, they were: -

Fig. 2
1913: Iona Davey (from a supplement in the Surrey Times) (Sarah Sullivan)

- Mary Watts the sculptress who had created the Watts Memorial Chapel at Compton and was also involved with the creation of Postman's Park, the cloister in London to commemorate the heroic deeds of individuals. (Fig. 1)
- Julia Charlotte Chance (nee Strachey) who lived at nearby Orchards, Munstead and was also a sculptress producing ornamental sculptures for some of the gardens designed by Gertrude Jekyll.
- Margery Horne who was a painter and wife of the local Conservative and Unionist Party MP.
- Iona Davey who was secretary of the Women's Local Government Society and wife of the candidate for the local Liberal Party. (Fig. 2)

After discussing the form of the new memorial it was decided to bring in Gertrude Jekyll, the well known gardener and plant specialist. A common factor between all five women was their membership of the National Union of Women's Suffrage Society (NUWSS) known as *Suffragists* (not to be confused with the suffragettes). The President of the Godalming and District branch was Mary Watts, with Gertrude Jekyll as one of the Vice Presidents. Julia Chance, whose husband was also a Vice President of the Society, wrote books such as *Words to Working Women on Women's Suffrage* (1912) and letters to *The Times* newspaper. By 1914 there were over one hundred thousand members with 500 branches throughout the country, but it was not until 1928 that all women were given the vote.

During the time whilst a location was sought for the Phillips Memorial, the NUWSS was involved in the Suffragist Pilgrimage of 1913. This was well documented by *The Times, Surrey Advertiser and Surrey Times,* although the event is now largely lost in the mists of time and overlooked by many historians. It was thought that a Women's Pilgrimage to London would demonstrate to Parliament how many women wanted the vote. NUWSS members from all over the UK spent nearly six weeks marching on London and holding a series of meetings en route. An estimated 50,000 women reached Hyde Park on 26th July. The meetings held on the way were not always peaceful with the proceedings at Farnham being pelted with eggs and police intervention being required at Guildford. The Pilgrims from the Portsmouth contingency made their way from Milford, escorted by the Godalming Town Band, to Ockford House for tea. (Fig. 3)

FIG. 3
1913, WOMEN MARCHING WITH GODALMING BANNER (GODALMING MUSEUM ARCHIVE)

'The arrival in Godalming excited a good deal of interest. About a hundred pilgrims were in the procession with banners and others followed in various vehicles'. These bicycles, traps and motor cars were decorated with flowers and ribbons in the suffrage colours of purple, white and green (purple standing for dignity, white for purity and green for hope).

Fig. 4
Banner designed by Gertrude Jekyll (Godalming Museum archive)

'Next to the band came the beautifully worked banner of the Godalming branch of the society, which was designed by Miss Jekyll'. (Fig. 4) In the evening a talk was held at the Borough Hall but due to limited space there was an outside or overflow meeting, at which Iona Davey was one of the principal speakers. Following collection and distribution of literature about women's enfranchisement the marchers were given accommodation by a number of prominent Godalming women, including Julia Chance and Iona Davey.

The five ladies on the Phillips Memorial Committee chose the form of a cloister, to be enclosed at night with solid gates and to have the appearance of old farm buildings. These *'buildings that have the merit and beauty of a simple aim and dignity that comes from the use of local material in the excellent traditional way of the country. Such a building as we advise could not without waste of space stand square with the road.'* It was planned to feature seats sheltered from wind and rain with a small garden in the midst.

The park was to be useful and *The Times* newspaper in approving the scheme pointed out that a memorial is no memorial if everyone forgets it as soon as it is set up. 'The cloister will be used. Its purpose will consequently be remembered. The name and story of Phillips will be kept alive'.

Horticulture, landscaping and the creation of public parks were subjects in which women excelled in this period and these five Phillips Memorial committee members all played their part. Julia Chance and Margery Horne had created exceptionally fine personal gardens with the aid of Gertrude Jekyll. The Chances had built a new home at Orchards, Busbridge and the Hornes had built Hall Place, Shackleford (now Aldro School) and Tigbourne Court, Witley for which Miss Jekyll supplied plans, layouts and plants. Mary Watts writes of a visit to Munstead Wood, home of Gertrude Jekyll, *'We drove to see Miss Jekyll in the afternoon, her polyanthus garden in full flower – so far no Jaune Desprez & Gloire de Dijon out on her walls – a peony Tree in full blossom – gentians too with their heavenly blue – in one patch – she has tried them in many places they all failed but there. She cannot tell me why a Banksia should not flower'*. Records

Fig. 5
1914: Gertrude Jekyll, Margery Horne and Mary Watts at the unveiling of the Phillips Memorial (John Young collection)

do not exist for Iona Davey's garden at Ockford House, Godalming (now known as Inn on the Lake) but there are certain elements that survive such as the massive Bargate stone wall with arched opening set within and the use of a tiled drip course that accentuates the arch and shows the Arts and Crafts influence.

The debate and struggle to find a suitable location for the memorial gardens continued in the local press on a weekly basis. It was the *Surrey Advertiser*, under the pseudonym *The Idler*, that first suggested the Glebe land where the Cloister was eventually to be placed, although it took many months before this was recognised as being a good solution and the land was purchased. Work on clearing the site adjacent to the church started, in late November 1913. The construction of the Cloister commencing in the winter months was probably not ideal and in February there were reports of the Lammas lands flooding extensively; but come what may this memorial park was to be finished in time for 15th April 1914.

During the research for this book a film by the local cinema owner, Mr W. G. Fudger came to light in the archives of the Screen Archive South East, University of Brighton. Entitled 'The Path of Duty was the Path to Glory', this very short, silent clip, taken on 15th April 1914, shows the women at the unveiling of the stone tablet. We see Gertrude Jekyll, Julia Chance, Margery Horne and Mary Watts seated on wooden chairs beneath the tablet. They are alongside the Mayor of Godalming, Harry Colpus with Alderman, Ernest Bridger and High Sheriff of Surrey, John St Loe Strachey facing a packed audience of over three hundred gathered to pay their respects. The fifth woman on the Phillips Memorial Committee was missing from this gathering as she was abroad with her husband. Although this is a very short film clip, it captures the cold wind blowing across the Lammas lands and the formal solemnity of the occasion. Archival footage brings the past to life and captures the atmosphere of a time that would otherwise just be lost to history books. (Fig. 5)

Gertrude Jekyll was to return to this park on many occasions and her planting notebooks and archives show later work there. Photographic archives of the Memorial Cloister also depict how successful some of the planting became, almost transforming the area into a romantic ruin shrouded in Virginia Creeper with tall grasses growing rampantly around the central pond.

GERTRUDE JEKYLL AND THE PHILLIPS MEMORIAL

Michael Tooley

Gertrude Jekyll returned to Surrey with her mother in 1878, a year after the death of her father in Wargrave-on-the-Hill, Berkshire. From 1878 until 1895, when her mother died, she was involved in laying out and planting the grounds of Munstead (House) in Busbridge. She had already acquired the ground to the north, in 1883, known as 'O.S.' (Other Side) and had begun to plant it up well before she met the young Edwin Lutyens in 1889. From 1878 until her death in 1932 she contributed to the life of the community both at Busbridge Church, which she helped to decorate at the time of the festivals of Christmas and Easter, and in Godalming and adjoining local areas designing the gardens and plantings for public buildings and memorials of which the Phillips Memorial was one of several. She also designed plantings for: in 1903 the Hughes Memorial Church (Wesleyan) Godalming; 1910 St. Edmunds Catholic Church, Godalming; 1914 Godalming Police Station; 1915 King's Arms Hotel; 1920 War Memorial at Compton, and in 1922 the War Memorial at Busbridge.

Her local work was part of a wider pattern of activity. By the late 1880s, she had already designed about sixteen gardens for friends and clients, written over 200 articles for the gardening magazines, such as *The Garden* and *Gardening Illustrated* founded by William Robinson and contributed a long section on colour in the flower garden in the first edition of the book he wrote on *The English Flower Garden* in 1883. She had also engaged in her first public works commission at Wargrave, for the Rev. Greville Phillimore in 1870 by designing a building over a spring, painting the tympanum showing Rebekah at the well of Nahor, and arranging the planting on the slope behind the building. This was the first of over twenty-five public works in which Miss Jekyll was engaged during her life and included work for the National Trust (Hydon Heath in 1915 and Ide Hill in 1921), King Edward VII Sanatorium in Midhurst in 1908, the War Graves Commission from 1918 and the Phillips Memorial from 1912. She designed over 400 gardens in Great Britain, Ireland, France, Germany, Yugoslavia and the United States of America (Tooley 2004, Tooley and Arnander 1995), and was engaged in six design commissions whilst working on the plans for the Phillips Memorial.

FIG. 1
THE THUNDERHOUSE, MUNSTEAD WOOD
(DRAWING SARAH SULLIVAN)

The Phillips Memorial Committee was set up by Godalming Borough Council shortly after the RMS *Titanic* sank on 15th April 1912 and reported on 9th

November, having sought the advice of Miss Jekyll on the form that the Memorial should take and how the ground should be laid out. At Miss Jekyll's behest, her friend, the architect Thackeray Turner of Westbrook, Godalming was involved. Miss Jekyll and Sir Lawrence Weaver were to write about the garden at Westbrook in *Gardens for Small Country Houses*. The Report included verbatim a forthright and practical letter from Miss Jekyll in which she outlined their scheme. She dismissed the initial proposal of a drinking fountain and portrait medallion on the grounds of cost and being of the opinion that an effigy of a young man in bronze or marble was rarely satisfactory.

What Jekyll and Turner proposed was a rectangular cloister 120 feet square, each corner occupied by a pavilion. A square garden in the centre of the cloister comprised four L-shaped beds enclosing a circular fountain basin or bed. The L-shaped beds would be planted up with low shrubs such as *Berberis* and Rhododendrons such as *R. myrtifolium*. An unbroken cloister would run along three sides and on the fourth opposite the entrance, where Bridge Road runs into the Meadrow, would be an arcade providing views south-east towards Unstead, the landscape value of which Miss Jekyll extolled.To the north-east of the Memorial she proposed a children's playground that ran along the back of the properties on the south side of the Meadrow. She wrote that a small stream, known as Hell Ditch, should be diverted and made shallower and thereby safer for children.

Miss Jekyll showed the same concern for the privacy of the landowners living adjacent to the proposed development as she showed to her northern neighbour at Munstead Wood;

FIG. 2
MISS JEKYLL PROPOSED LOCATING THE MEMORIAL OPPOSITE GODALMING TECHNICAL INSTITUTE (JOHN YOUNG COLLECTION)

when the triangular belvedere, designed by Lutyens, was built, she had the north opening that looked down onto her neighbour's garden, closed by wooden shuttering. (Fig. 1) At the time of the meeting of the Town Council on 9th November 1912 when the Report of the Phillips Memorial Committee was tabled, the town clerk, Mr.T.P.Whatley, reported that none of the owners possessing rights to the Lammas Land upon which both the Memorial and the children's playground were to be located had objected. According to Miss Jekyll the adjoining owner, Mr. W.I.Nash, a retired councillor, who initially had consented to the building, had also offered an independent access to the meadow, which she urged the Council to accept.

However, notwithstanding Miss Jekyll's concern and recommendation that, 'we feel it to be essential that the adjoining owner, Mr.Nash, should not suffer any annoyance from our scheme, and have therefore shown a separate entrance which gives access to the back of his premises,' on 1st August 1912 Mr. Nash objected. The Council attempted to meet his objections by proposing to erect a 6ft. high fence 30 ft. from his property 'to keep the children that distance away.' However, six months later he withdrew his consent, argued that the Memorial as proposed would obstruct his view and the playground would be 'an intolerable nuisance,' and he would seek compensation if the Council pressed ahead with this project. He was prepared to agree to the original proposal of a statue and seating on the present site and offered 1½ acres of land in the Meadrow facing the entrance to Llanway Road for the Memorial as proposed by Miss Jekyll and Mr. Thackeray Turner.

Miss Jekyll was invited to comment on its suitability. Her letter to the Council was reproduced in the *Surrey Advertiser* on 14th June 1913. She expressed disappointment that the site was remote and could not play any part in the social life of the town. The Town Band could not be expected to attend at this distance from the centre. There were no other buildings nearby, and the Memorial would have to be redesigned; the site was outside the '10 mile limit for motors' and there was no footpath, making access dangerous for children; the views to Langham were inferior to those from the Memorial site on the Lammas land. She opined that a better location would be opposite the Technical School, where Bridge Road runs in to Chalk Road. (Fig. 2) From a landscape point of view, the open arches, 'would command not only a superb view of meadow, stream and wooded background, with the church as a central object, but as they would face nearly west the view would embrace all the fine effects of sunset scenes, with interesting change of aspect of the frequent occasions when the meadow flooded.'

Her view did not prevail and the Phillips Memorial was built on Glebe Land immediately south of the River Wey and north of the parish Church of St. Peter and St. Paul adjacent to the road to Charterhouse. The design was little changed from the original proposal. It was an enclosed court with a covered cloister on three sides and a screen of brick on the east side with seven arches, the three central ones blind. (Fig. 3) The open arches afford views east over meadowland adjacent to the river Wey. A raised octagonal tank with eight borders 9ft.8in. by 2ft.6in. occupied the centre of the Memorial. (Fig. 4) There were two borders flanking the memorial tablet some 20ft. long and 4ft. deep. (Fig. 5) Outside the Memorial there was a south border facing the Church and shrub plantings to the west between the Memorial and the road. Miss Jekyll laid out the ground adjoining the Memorial and this comprised a Bowling Green and Quoits Alley to the south and shrub-enclosed lawns to the east. Access to the Cloister was by way of a stepped path from the west and a level entry to the south cloister from the footpath running west to east.

Miss Jekyll provided three detailed planting plans and most of the plants from her nursery at Munstead Wood. (Fig. 6)

Within the Cloister the evergreens provided a foil for the flower colours. The east borders beneath and adjacent to the open arches contained at the back laurustinus (*Viburnum tinus*) with pink buds and white flowers from October to March, the aromatic – leaved bay (*Laurus nobilis*), Wistarias and pink flowered kalmias. In front were groups of the white-flowered *Veronica Traversii* (*Hebe brachysiphon*) and two species of rhododendrons with rose-coloured flowers – *R. myrtifolium* and *R. ferrugineum*. These were interspersed with acanthus, irises (*Iris sibirica* blue and white, the white flowered *Iris ochroleuca*), white flowered *Campanula persicifolia* and *Phlox* 'Avalanche' (*Phlox maculata* Avalanche = 'Schneelawine'). At the front were drifts of *Veronica* (*Hebe*) *buxifolia*. Around the tank were planted low growing evergreen shrubs, such as *Rhododendron* x *praecox* that flowers in February and March, skimmias and *Veronica* (*Hebe*) *buxifolia*. Much use was made of Male Ferns and Lady Ferns, and herbaceous plants included aquilegias, bergenias, hostas and the Welsh Poppy *Meconopsis cambrica*. There were drifts of blue-flowered *Geranium grandiflorum*, London Pride, *Nordmannia* (*Trachystemon*) and *Omphalodes verna*.

The west wall of the Cloister facing the road was planted up with 3 *Clematis montana*, 4 Virginia Creepers, and 6 clumps of tree lupines. Flanking the entrance were two *Azara dentata* from Chile.

FIG. 3
1914, CLOISTER: SOUTH EAST VIEW (JOHN YOUNG COLLECTION)

Fig. 4
1914, Cloister: raised tank (John Young collection)

The awkward-shaped pieces of ground rising to the road to the west of the Memorial were to be planted up with bamboos, *Veronica* (*Hebe*) *brachysiphon*, Water Elder and Guelder Rose, snowberries, kerrias, jasmine, Portuguese laurel and berberis. Between the Memorial and the River Wey she proposed to use two exotic plants that are now proscribed – the giant hogweed and balsam. She was particularly fond of *Heracleum mantegazzianum* from Abkhasia at the foot of the Caucasus, which she described thus: 'this grand plant is not only larger in all its parts but in spite of its size has an appearance of greater refinement

Fig. 5
1914, Cloister: memorial tablet (John Young collection)

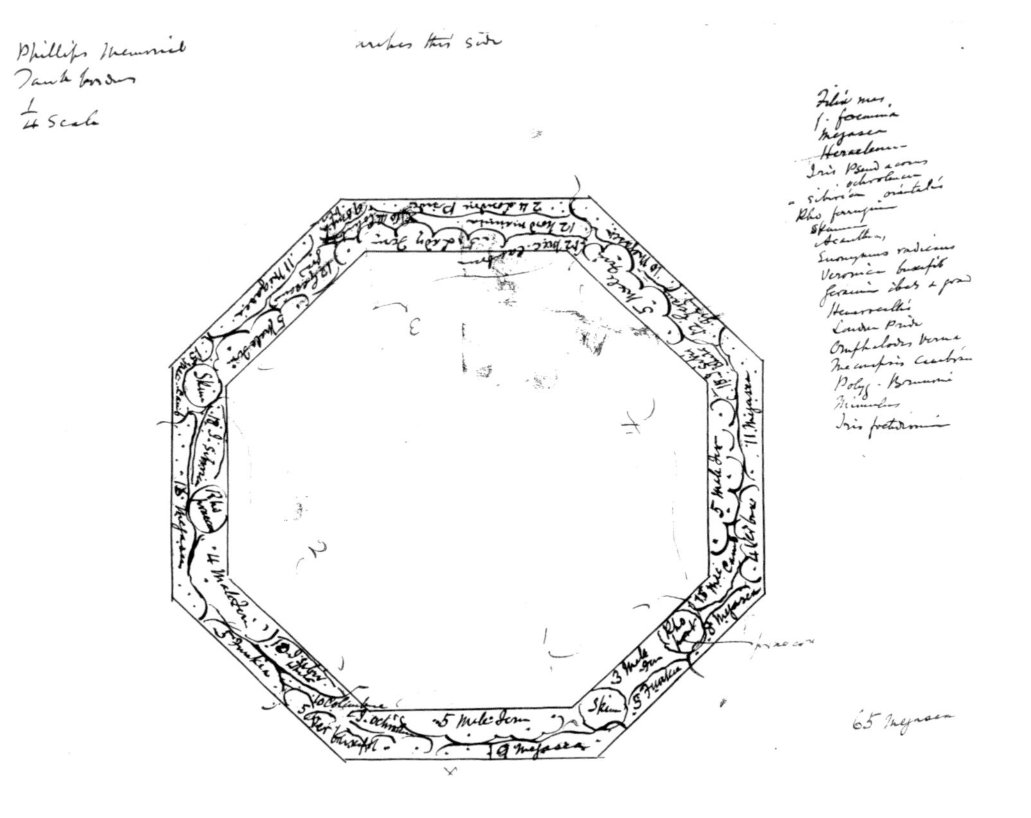

Fig. 6
Planting plan for tank border (Gertrude Jekyll Collection (1955-1), Environmental Design Archives, University of California, Berkeley)

and beauty' than the smaller *H. giganteum*. She attributed to it a 'mien of a specially proud and sumptuous plant,' and proceeded to introduce it in most of the planting plans for woodland and waterside gardens throughout the country. Of balsam, for which she supplied seedlings of the annual white *Impatiens glandulifera* 'sold under the doubtfully correct name *I.* Roylei' she wrote in 1924, ' it will grow from seven to eight feet high, branching wide with its fragrant pure white flowers of true balsam form suspended in clusters from the ends of the branches and axillary shoots.' This plant has followed most rivers throughout England in the past eighty years being invasive and difficult to eradicate.

There are two 30ft. long south borders flanking the entrance path in the south cloister. The border is backed by a wall against which were planted escallonia, probably *E.Philippiana* (*E.virgata*), ceanothus, senecio, Japanese honeysuckle and *Spiraea Lindleyana* (*Sorbaria tomentosa*). The full stop at the end of each border was the evergreen dark leaved Japanese privet, *Ligustrum japonicum*. White flowered shrubs and herbaceous plants - *Veronica* (*Hebe*) *buxifolia*, *V.*(*H.*) *brachysiphon*, *Clematis flammula* and *Centranthus* - were a foil for the yellow tree lupines, blue lupines (probably the lupin she bred and was known as 'Munstead Blue'),

Fig. 7
1916, view of the Cloister interior (John Young collection)

Fig. 8
1920s view of the Cloister's matured planting (John Young collection)

Fig. 9
Cloister with Godalming church and showing the open pergola added in 1965. Drawing by Rosanna Tooley, (M. Tooley)

orange montbretia and day lilies and two fuchsias flanking the path into the cloister. (Fig. 7) and (Fig. 8).

Well over 700 plants for the Memorial gardens were supplied from Miss Jekyll's plant nursery at a cost of £17..5..0 (£17.25) and the balance (62 plants) bought in from nurseries at a cost of £4..11..8 (£4.57) and paid for by Miss Jekyll. The entry for the Phillips Memorial in the Account Book in Godalming Museum archive does not tally with the plans for the borders. For example, on the plan for the south border Miss Jekyll shows plantings of *Centranthus*, *Aconitum napellus* and fuchsia that do not appear in the account book lists. But, *Monarda* is not shown on the plan but appears in the account book.

Miss Jekyll's overall plan and plantings for the Phillips Memorial and the adjoining Park, although not at the site she would have preferred further east, considerably enhanced the building and integrated it with the water meadows to the north and east and the Church and War Memorial to the south. (Fig. 9) Her description of the Memorial some ten years later in *Gardening Illustrated* reveals none of the problems that beset the project and both the photograph and factual report endorse the successful collaboration between Miss Jekyll and Mr. Thackeray Turner.

Acknowledgement: it is a pleasure to acknowledge the help that the late Mrs Joan Charman gave me in the library of Godalming Museum from 1990 until 1998. Relevant books and letters appeared on the table as I worked, with information about Miss Jekyll's contributions to the life of Godalming qualifying the widely held view of her reclusiveness. I am grateful to Sarah Sullivan for providing copies of newspaper articles.

HUGH THACKERAY TURNER – ARCHITECT AND CONSERVATIONIST

Robin Stannard

To design the new Memorial Cloister, Gertrude Jekyll turned to her friend, the Godalming architect, Hugh Thackeray Turner. (Fig. 1) Modest and reticent in character, he also thrived on debate and controversy, aspects of his character that he often used in his tireless championing of building preservation and protection of the natural environment.

Turner was born in 1853, in Foxhearth, Essex, the son of a country vicar. He was one of seven children of a creative and artistic family. His elder brother Hawes, was a painter, who eventually became the keeper of the National Gallery, whilst his younger brother Laurence, trained as an architect, but became a notable Arts and Crafts carver. Turner was educated at Newbury Grammar School and then served his articles with Sir Giles Gilbert Scott, one of the nineteenth century's most successful architects and a leading exponent of the Gothic Revival. After Scott's death, in 1877, Turner became the chief assistant to his son, the gifted architect, George Gilbert Scott junior, whose tragic decent into madness led him to be eventually incarcerated in the Royal Bethlem Hospital.

FIG. 1
HUGH THACKERAY TURNER SEATED ON A BENCH IN WESTBROOK'S GARDEN.
(PHOTOGRAPH: JUSTINE VOSIN)

The turning point in Turner's life occurred in 1883, when he was offered the position of Secretary of the Society for the Protection of Ancient Buildings. The society had been founded by William Morris in 1877, with the aim of protecting medieval churches from overzealous restorations and demolition by Gothic Revival architects, such as Sir George Gilbert Scott. Turner held the position as Secretary for 29 years, during which time the society changed from being a reactionary protest group into a respected, professionally run, conservation body. His appointment drew him into the artistic circle which centred on William Morris, together with a group of architects who would form the core of the evolving Arts and Crafts Movement. These architects included Philip Webb, William Lethaby, Charles Ashbee, Ernest Gimson, Sidney Barnsley, George Jack, Alfred Powell and Detmar Blow. In 1885, Turner entered into partnership with the SPAB architect, Eustace

FIG. 2
WYCLIFFE BUILDING, GUILDFORD, DESIGNED BY TURNER IN 1894. (PHOTOGRAPH: ROBIN STANNARD)

Balfour, brother of the future Prime Minister. The practice, named Balfour Turner, designed many buildings in Mayfair, bringing the architecture of the Arts and Crafts Movement into the heart of the capital.

It was through the SPAB that Turner made the acquaintance of the wealthy stockbroker Thomas Wilde Powell. Powell had recently moved to Guildford and lived in a country house designed by Norman Shaw, called Piccards Rough. Powell used his wealth to support his interests in art, design, and conservation of the natural and built environments. He was also a dedicated philanthropist, supporting many local causes, particularly those relating to education and the provision of housing for the poor. In 1888, Turner married Powell's daughter Mary, a notable Arts and Crafts embroideress. Mary would later, in 1907, found the Women's Guild of Crafts, with William Morris's daughter, May Morris. The couple lived in Gower Street, London, and had three daughters, Marjorie, Ruth and Mildred. They often spent their weekends in Gomshall, where Turner would explore the local countryside on his bicycle, sketching nature and vernacular buildings. With the coming of the railways, South West Surrey had become the ideal place to live for people who wished to escape the unhealthy environment of London. Turner was aware of the threat to the area posed by new building development and was particularly concerned about protecting Guildford's historic High Street. Through his position as Secretary of the SPAB, Turner encouraged the formation, in 1896, of the Old Guildford Society.

FIG. 3
THE COURT, GUILDFORD, DESIGNED BY TURNER IN 1901. (PHOTOGRAPH: SARAH SULLIVAN)

In 1894, Powell commissioned Turner to design the Wycliffe Building, Guildford. (Fig. 2) This was a successful experiment in the provision of working class housing, and an important example of Arts and Crafts architecture. Through various members of the Powell family, Turner undertook a number of other commissions in Surrey. In Guildford, he designed Millmead Cottage, Piccards Cottage and The Court (Fig. 3) whilst in Dorking he designed the Powell Corderoy School and Goodwyns Place (Fig. 4).

The death of Thomas Wilde Powell, in 1897, resulted in Turner's wife receiving a substantial inheritance. This enabled the couple to move permanently from London and to build a home in the country. The site chosen, in Godalming, was dramatically situated on a hill, facing north towards the spires of Charterhouse and the River Wey Valley. The house, called Westbrook, was completed in 1900. It is an outstanding example of Arts and Crafts design and is one of Turner's most important works. In typical Arts and Crafts fashion, the house was built of local Bargate stone, dug from the site. The house successfully merged local vernacular materials and building traditions with the latest Edwardian technology. This included electric lighting, warm air heating, concrete floors and structural steel roof construction. Importantly, the house and garden were designed as a single entity with internal planning of the house carried through to the layout of the garden. The design of the garden was carefully planned to give a variety of spaces, including intimate formal walled gardens, organically designed wild garden areas and a large expanse of lawn giving distant vistas. Within the walled gardens, seats were placed, with roofs over, to give protection from summer sun and rain. One of the most important areas was an octagonal shaped sunken garden and pool. The planting layout for this was designed by family friend, Gertrude Jekyll. (Fig. 5) Many elements in the design of the garden can be seen as an inspiration for the design of the Phillips Memorial Cloister. Gertrude Jekyll was lavish in her praise of the garden dedicating a chapter to it in *Gardens for Small Country Houses.* The house and garden were also featured twice in *Country Life* Magazine.

FIG. 4
GOODWYNS PLACE, DORKING, DESIGNED BY TURNER IN 1902. (PHOTOGRAPH: ROBIN STANNARD)

Following the example of Thomas Wilde Powell, Turner immediately became involved with the local community and local causes. The first campaign was an attempt to save the medieval Town Bridge in Guildford, which had been damaged by a storm in 1900. Despite considerable efforts, the bridge was replaced by a single span cast iron bridge, which exists today. Turner's failure to save the bridge resulted in him forming the West Surrey Society,

Fig. 5
Westbrook, Godalming. View of sunken garden with raised octagonal lily pond, similar to Phillips Memorial, c1933. (Photograph: Justine Vosin)

in 1907, which was dedicated to the protection of local historic buildings and landscape. He was more successful in his efforts to save Eashing Bridge, which he restored in 1902. Turner campaigned vigorously to save the Pepper Pot in Godalming from demolition, eventually taking out a seven year lease to prevent its destruction. (Fig. 6) Mary Thackeray Turner was also active in her support of conservation, contributing £500 to save the Devils Punch Bowl for the National Trust. In 1906 she bought 240 acres of land, at Witley and Milford Commons, to prevent them being developed. The couple were heavily involved with the local Godalming community, buying the old Liberal Club in Bridge Street, and converting it into a men's club, which contained a billiards room, library, and a bar selling non-alcoholic drinks. At Westbrook, a kitchen was built in the garden where Mary and her daughters taught cookery to the Girls Club. Turner was a manager of Moss Lane School, where he is reputed to have played his flute to the children.

Turner was also a noted ceramics painter, an interest which he had begun in the late 1870's and continued until the 1930's when poor eyesight prevented him continuing. Locally, both Godalming and Guildford Museums have examples of his work. (Fig. 7)

In 1907, Turner suffered a great tragedy when his wife suddenly died of pneumonia leaving him with three teenage daughters to bring up. This situation was exacerbated by the worsening health of his partner, Eustace Balfour, who died in 1911. Turner was badly affected by this and as a result resigned from his position as Secretary of the SPAB. To take his place, he recommended that the SPAB appoint a young conservation architect, Albert Reginald Powys (1881-1936), who had previously worked for Arts and Crafts Architect, Walter Cave and gained much hands-on building conservation experience with SPAB architect, William Weir. A few months later Turner took Powys into partnership in his practice, Balfour Turner and then assumed a more strategic role as Chairman of the SPAB Committee, a position he held until his death. It was against this background that in 1912, Turner was asked by Gertrude Jekyll to collaborate in the design of the Phillips Memorial Cloister.

In 1913, Turner made the acquaintance of a young teacher called George Mallory. Already a noted mountain climber, in 1914 Mallory married Turner's daughter Ruth. Tragically he lost his life in 1924, during his ill-fated attempt to climb Everest.

Turner continued to serve the local community, buying an old barn in Charterhouse Road which he converted for community use; it eventually became the Headquarters of the Godalming Boy Scouts. In 1921, Turner and his daughters gave Witley and Milford Commons to the National Trust and he also saved ancient timber framed cottages in Eashing and Ockford Road, Godalming, which were gifted to the National Trust. Turner continued his work for the SPAB and National Trust, until his death in 1937. His achievements, in both architecture and conservation, being summed up in the words of his memorial in the Church of St. Peter and St. Paul, Godalming:

An architect and an artist in the craft of building, who devoted most of his life to the saving and repairing of the ancient buildings of England, and to the preservation of the beauties of the countryside.

FIG. 6
GODALMING MARKET HOUSE: KNOWN AS 'THE PEPPER BOX', LATER THE 'PEPPER POT'- SAVED FROM DEMOLITION BY THACKERAY TURNER. BEYOND: THE POST OFFICE WHERE JACK PHILLIPS WORKED.
IN THE FOREGROUND: THE OLD COBBLES, REUSED AT THE MEMORIAL CLOISTER (JOHN YOUNG COLLECTION)

FIG. 7
MOON FLASK DECORATED BY HUGH THACKERAY TURNER (PHOTOGRAPH: GUILDFORD MUSEUM)

THE ARCHITECTURAL DESIGN OF THE PHILLIPS MEMORIAL

Robin Stannard

The Phillips Memorial Cloister is a fine example of both Arts and Crafts architecture and Hugh Thackeray Turner's approach to design. It was originally built as an enclosed courtyard, surrounding an octagonal pool. (Fig. 1) During the 1960s the south wall was removed and replaced with an open pergola. Although this has changed the introspective nature of the cloister, in other respects, the design remains much the same as when it was originally built.

The design of the Cloister was influenced by the shared interest of Gertrude Jekyll and Turner in the vernacular buildings of South West Surrey, Jekyll stating:

> *'The style of the building that we contemplate is that of the older of the local farm buildings – buildings that have the merit and beauty of a simple aim and dignity that comes of the use of local material in the excellent traditional way of the county'.*

Although the design of the Phillips Memorial Cloister was influenced by traditional local farm buildings, it is not a copy, but is a subtle original design. Turner, then aged 59, drew on many years experience as an architect and on his work inspecting historic buildings on

Fig. 1
Interior of Cloister looking towards lily tank. (Photograph: Sarah Sullivan)

PLAN OF THE PHILLIPS MEMORIAL CLOISTER

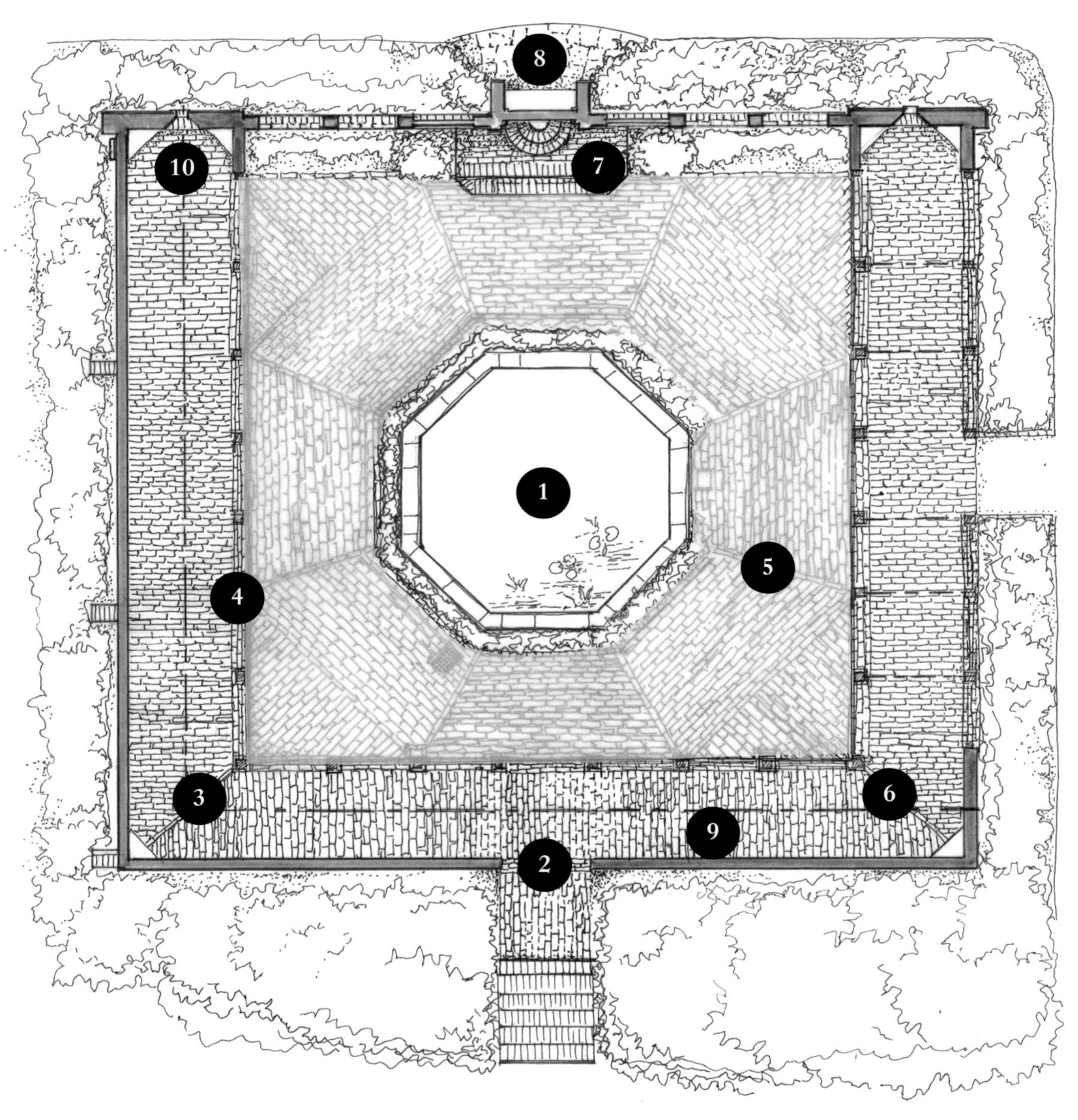

Figure numbers refer to the illustrations within the essay
Drawing by Sarah Sullivan

FIG. 2
ENTRANCE TO THE CLOISTER WITH SUBTLY RAISED CURVE IN THE EAVES LINE AND PLAIN CLAY TILED QUOINS TO THE SIDES. (PHOTOGRAPH: SARAH SULLIVAN)

FIG. 3
TRADITIONAL LACED VALLEY TILE DETAIL AT ROOF JUNCTION (PHOTOGRAPH: SARAH SULLIVAN)

behalf of the Society for the Protection of Ancient Buildings. This gave him an intimate knowledge of traditional buildings, which had an important effect on his approach to architectural design. As well as undertaking the design, Turner also supervised the construction, and although the Cloister appears as though it had developed over time, in fact every detail is meticulously considered and executed. At first sight the building appears symmetrical, however each of the external walls of the Cloister is different. This was a deliberate design decision influenced by the way traditional buildings gradually evolve to suit different needs. It was a design technique that Turner used on many of his buildings, including Goodwyns Place, Dorking and Lygon Place, London. Central to Turner's approach to design was the belief that buildings should be honest and have integrity; attributes which were also at the heart of his own character.

The orientation of the Cloister is reminiscent of a church with the memorial placed to the east and the main entrance at the west. In a typical Art and Crafts manner the entrance is deliberately understated, only being denoted by a gently raised curve in the eaves line. (Fig. 2) A second entrance was located in the north wall, which was removed in the 1960s. The central octagonal pool was originally surrounded on three sides by a covered cloister, to enable it to be used in all seasons. Within the Cloister, large oak seats were provided, designed by Turner, which were based on a pattern which he had used at Westbrook. Jekyll describes the intended use of the Cloister as follows:

'There would be seats under the cloister, and, according to wind and weather, there would always be some portion providing protection. It would be of special utility to nurses and young children providing a safe place for air and exercise in threatening weather, as well as of pleasure to the many who would be glad of such a spot of reposeful beauty and harbourage.'

FIG. 4
OAK POST WITH SUBTLY DESIGNED CLASSICAL ENTASIS. (PHOTOGRAPH: SARAH SULLIVAN)

A further covered seat was provided externally on the east elevation overlooking a proposed children's playground.

The roof to the Cloister is covered in handmade plain clay tiles, supported on oak posts with gently tapering entasis following the pattern of classical columns. (Fig. 4) The smooth plastered ceiling features a typical Turner deep cove detail. Paving to the Cloister consists of cobbles which were salvaged from Godalming High Street. (Fig. 5)

The east elevation consists of open and blind arcading built in brickwork, of considerable height, unrestrained by buttressing and slender in thickness, showing Turner's confidence and experience as an architect. The two pairs of open arcades is a theme Turner had used on other buildings, and here they are used to frame views of the sky and distant landscape. (Fig. 6)

FIG. 5
COBBLE STONES RECLAIMED FROM GODALMING HIGH STREET (PHOTOGRAPH: SARAH SULLIVAN)

Fig. 6
Interior of cloister with arcading and open pergola. (Photograph: Sarah Sullivan

The focal point on the east wall is the memorial plaque to Jack Phillips. (Fig. 7) This was carved in local Bargate stone by Turner's brother Laurence. Laurence had originally trained as an architect, but is best known as a carver, working in stone, wood and plaster. He worked for many leading architects including George Bodley, William Weir, Walter Tapper, Curtis Green, Sir Herbert Baker and Philip Webb. Notably with Baker he produced carved interior panels for South Africa House and for Webb he carved William Morris's tombstone at Kelmscott. He is also credited with undertaking the timber work to the façade of Liberty's well known shop in London. Laurence is now less well known than his contemporary Eric Gill, having preferred to perfect the craft of carving, rather than undertake the path of more radical design. As a craftsman he was a noted teacher and one of his more important pupils was the architect George Jack, who later designed and carved the reredos for the Church of St Peter and St Paul, Godalming. Stuart Gray (A. Stuart Grey, *Edwardian Architecture. A Biographical Dictionary* (Wordsworth Editions 1988) p357) stated, 'Turner taught by his drawings and held exacting standards, preferring to lose money rather than economise on his work'. Laurence collaborated extensively with his brother, outstanding examples being the exquisite floral stone carving surrounding the front door to a house in Green Street, London and plaster carving at Westbrook, Godalming. The floral border to

Fig. 7
Memorial tablet to Jack Phillips, carved by Laurence Turner. (Photograph: Sarah Sullivan)

the Phillips Memorial plaque is a typical example of Laurence's work. Below this, Laurence carved an elegantly shaped stone water fountain donated by the Postal Telegraph Clerks Association.

In typical Art and Crafts manner the Cloister is built of traditional local materials. The walls are built from purple/brown multi–coloured, local, handmade bricks. During the Victorian period these bricks had been shunned as inferior rustic bricks and often were only used when covered by other materials, or used for the less important rear elevations. The Victorians preferred to use higher quality red bricks for important elevations, but such bricks were sometimes transported by train from other areas. Turner chose to use the locally made brick honestly and to great decorative

Fig. 8
Detail of contrasting orange tiles used to emulate classical modilions to the external shelter (photograph: Sarah Sullivan)

Fig. 9
Brickwork in Irregular Bond laid in lime mortar (photograph: Sarah Sullivan)

Fig. 10
Internal view of canted arrow slit opening, showing Turner's use of decorative tiles to head and sill (photograph: Sarah Sullivan)

effect in combination with contrasting orange clay tiles. This practice was one that Turner used on many of his buildings, directly influenced by his conservation work for the SPAB, which advocated the use of clay tiles for the repair of medieval stone buildings, to show with honesty where they had been repaired. An example of this technique can be seen on the walls of Guildford Castle. Good examples of Turner's decorative use of clay tiles may be found at Goodwyns Place, Dorking and Shottendane House, Margate.

The brickwork is laid in lime mortar arranged in irregular bond. (Fig. 9) The use of such a brick bond is highly unusual for an architect of this period, being commonly associated with early medieval brickwork, or crude farm buildings. The latter association is very appropriate to the design intention of the Cloister. Brick bonds more generally used at this period were based on variations of either Flemish or English bond, the advantage of these being their consistency of strength and avoidance of weak points caused by straight joints. The successful use of irregular bond brickwork relies on the skill and supervision of the bricklayers to ensure strength and avoid straight joints. At the Phillips Memorial the quality of the brickwork is consistently high, testifying to the skill of the bricklayers and Turner's supervision. The advantage of the use of irregular bond brickwork is that it gives a much softer appearance than the harsh rhythm of regularly bonded brickwork.

Turner took delight in small details, a typical example being the arrow slot openings at the east end of the cloisters. (Fig. 10) These show his romantic interest in medieval buildings and were also doubtlessly designed for the pleasure they would bring to small children, stimulating their minds to imagine ancient castles and knights in armour.

THE PHILLIPS MEMORIAL PARK – LOOKING AFTER THE HERITAGE

Nick Baxter and Lizzie Noble

'Let us hope that Phillips' example and Phillips' memory may become a part, as it were, of the building – a spell to bind the spirits of those who enter here'

High Sheriff, Mr J. St Leo Strachey at the opening ceremony, 15 April 1914

The heritage described

The Park dates back in part to 1913, when land was purchased for public enjoyment and exercise, and to provide a location for the Phillips Memorial Cloister. At this time, it was known as the Phillips Memorial Gardens. (Fig. 8) The Rectory Animal Pound was later given to the people of Godalming in 1933, by the Rector of the Church of Saint Peter and Saint Paul. Further parts of the site were purchased in 1952 and 1959. In 1966, the Municipal Borough of Godalming created the modern day park, by amalgamating the three separate but contiguous components of the present day park (known as the Phillips Memorial Gardens, the Burys and the Burys Youth Field) under the umbrella title, the Phillips Memorial Park. (Fig. 1)

The western end of the Park, including the Phillips Cloister, the Bowling Green, the Animal Pound and the War Memorial, lies within the Godalming Conservation Area (designated in March 1974, and extended in October 1984 and December 1989). In 1991, the Cloister was listed as grade II. Heritage features, covered by Waverley Borough Council's Local Plan under Policy HE10, include the Rectory Manor Animal Pound, the War Memorial and adjacent Bargate stone walls (Bargate stone is a calcified sandstone specific to this part of the Weald), as well as the Phillips Memorial Gardens themselves.

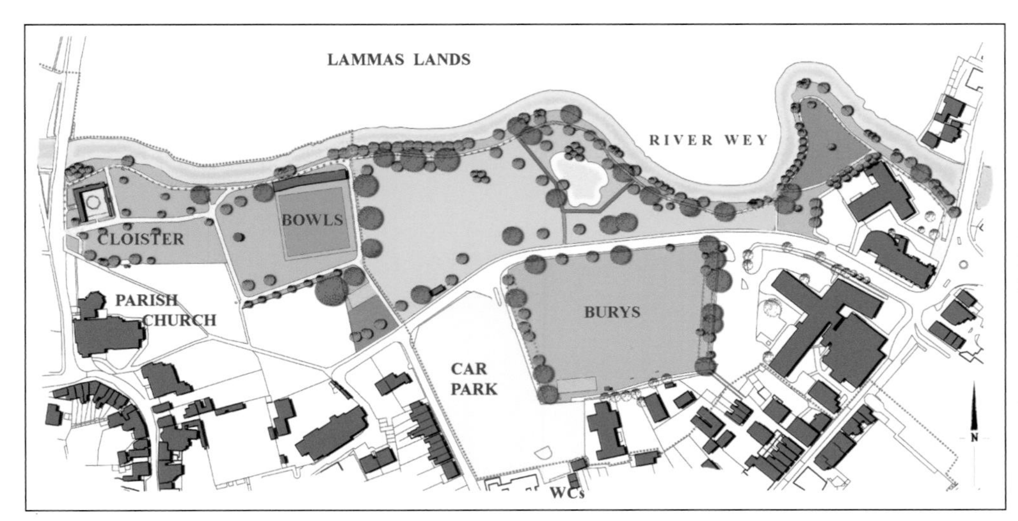

FIG. 1
MAP OF PHILLIPS MEMORIAL PARK (AFTER WAVERLEY BOROUGH COUNCIL).

Reproduced by permission of Ordnance Survey on behalf of HMSO. Ordnance Survey Licence number 100052562

Fig. 2
Sketch giving an impression of the proposed Cloister, from the Local Paper dated 7th December 1912

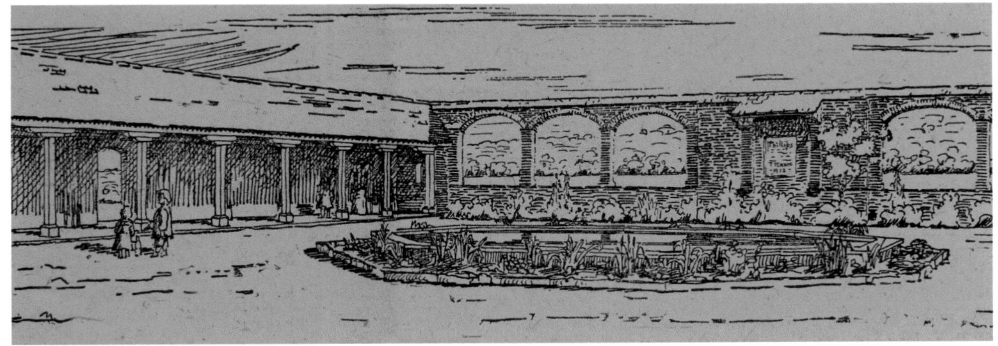

The modern day park is multi-faceted, and apart from the built heritage and provision for formal and informal recreation, also includes areas of semi-natural vegetation (such as the meadow area and the wet scrape) as well as areas of formal planting. It has an important landscape function, providing a screen between the two sides of the river valley, whilst also providing an important green corridor, together with the Lammas Lands, alongside the River Wey through the centre of Godalming. For these reasons its comparatively small size belies its huge intrinsic value, as it contributes significantly to the overall biodiversity, landscape, history and informal recreation value of the area. Successful management needs to take account of all these components of the heritage.

The first 50 years

Originally, the Cloister was seen as a place for nurses and their charges to perambulate, and also as a suitable venue for musical concerts. It was always intended that the Cloister should be seen as an integral part of the whole Park, and Jekyll included plans for quoits and bowls to be played outside. (Fig. 2)

The *Surrey Advertiser* reported, on 29th November 1913,

'The Development Committee of the Corporation are also formulating a scheme for the erection of a bathing place on the land which has been acquired and which abuts onto the river and for the laying out as a pleasure ground of the surplus land'.

This plan was dropped as the site was deemed unsuitable, owing to its boggy nature and the bathing area was relocated further downstream, to Catteshall.

The Cloister was the focal point of the garden. Jekyll designed the planting areas in and around the Cloister, plus around the ornamental pond, as previously described. She planted 2 bay trees (still in existence) either side of the (now disconnected) commemorative drinking fountain at one end of the Cloister. The fountain was donated by the Postal Telegraphic Clerks Association. Above the fountain, a memorial tablet is surmounted by the Godalming Borough arms bearing the following subscription:

S.O.S – This cloister is built in memory of John George Phillips, a native of this town, chief wireless telegraphist of the ill-fated RMS *Titanic*. He died at his post when the vessel foundered in mid-Atlantic on the 15th day of April, 1912.

FIG. 3
1900S, GODALMING CHURCH, SHOWING VIEW OF THE ANIMAL POUND AND LAND ON WHICH THE MEMORIAL WAS BUILT (JOHN YOUNG COLLECTION)

It seems that the committee also planned that the walls in the Cloister should be used for tablets celebrating other brave acts by local people (similar to the Watts-inspired Postman's Park in London); however, this never happened.

Only months after the opening, it became evident that the Park was being well appreciated. Children were observed playing regularly in the gardens, the Town Band had played in the Memorial and Jekyll continued to assist with planting bedding.

The Animal Pound (a reminder of the days when the area was grazed by free-roaming animals, and strays would have been impounded by the Lord of the Manor and only released when a fine had been paid) was thought to have been moved when the Cloister was built and the road widened in 1913/14. Close inspection of the various Ordnance Survey maps suggests that the main part of the Pound remains roughly where it has always stood, but the northern and southern walls have both been moved southwards towards the Church. (Fig. 3)

Post-World War I, in 1920, saw the revival of the Thackeray Turner and Jekyll partnership with the dedication of the Town War Memorial. (Fig. 4) Built within the Phillips Memorial Garden, the War Memorial was designed by A R Powys, a contemporary and architectural partner of Thakeray Turner. Two extant commemorative yew trees were planted either side by Jekyll and Thackeray Turner (who was then also a church warden). Around the same time, Jekyll was thanked publicly, in the *Surrey Advertiser*, by the then Town Mayor

FIG. 4
GODALMING WAR MEMORIAL (PHOTOGRAPH: NICK BAXTER)

for her help with planting. This may relate to the planting of the yew trees or there is evidence that it more likely related to the continued involvement of Jekyll in the maintenance of the formal gardens.

Following representations to the Municipal Council in 1923, the bowling green was constructed at a cost of £170. This was let to the Godalming Bowling Club in 1924, and has been home to the Godalming and Farncombe Bowling Club since the two clubs merged in 2005.

Human nature has changed little in a century, and the Cloister and Park have suffered from the occasional outbreak of anti-social behaviour. It is reported that sanitary conveniences were placed in the park in 1926 following a request from the public; thought to be situated between the Cloister and the river, they were however swiftly removed following vandalism.

The *West Surrey Outlook* reported in 1954 that the weeping willows in the Phillips Memorial Garden had reached their prime, and, in 1956, it was recommended that nine of the trees should be felled. This caused a great deal of upset to local residents who felt that the trees were one of the *special features of Godalming beauty*. Replanting and replacements were carried out to maintain this iconic aspect of the landscape.

The second 50 years

1965 proved a significant year for the Park. The Riverside Walk was constructed, with a surfaced footpath and the slopes down to the river edge were graded, the path forming a link between the Phillips Memorial Garden and The Burys. Following the construction of the Riverside Walk and The Burys road, the Municipal Borough of Godalming agreed, in 1966, that the Burys pleasure ground and the Phillips Memorial Garden should be treated as one, although the field lying to the south of the Burys and which formed part of the overall park, was still to be known as The Burys Youth Field.

In 1965, tiles on the roofs on the north and west sides of the Cloister were stripped and repairs carried out to prevent further damage to the ceiling caused by water penetrating from above. In the same year, the Cloister was changed dramatically. The southern wall was removed, apparently because of structural failure, but also to open the Cloister up, to reduce the likelihood of inappropriate use of the Cloister, which was the cause of some concern! In place of the solid wall, an oak pergola was erected to act as a support for the existing *Wisteria*. Unfortunately, closer inspection of the *Wisteria* showed most of it to be dead, and it was removed. The work called for consideration of the surrounding planting; before work began four poplars, located in the narrow space between the Borough Road and Cloister, were felled. Replanting of the *Wisteria* and other formal plant beds in the garden was also carried out in 1965.

Local Government re-organisation in 1974, led to the formation of Godalming Town Council and Waverley Borough Council. Ownership of the Park was then taken over by

Waverley Borough Council and managed by the Parks and Landscape Team, although two small strips of land adjacent to the two bridges at either end of the Park passed to the Town Council, supposedly to allow for snow from the town centre to be pushed into the river.

In 1986, due to considerable difficulty in maintaining the ornamental pond in a litter-free condition, it was decided by the WBC Leisure and Culture Committee that it would be converted into a shrub bed. As part of a community project, students from Charterhouse were involved with in-filling the pond with soil and it then had a comparatively short life acting as a large container for plants. A children's playground was installed in the Park in November 1989, and dog-proof fencing was erected around the outside in June 1996.

The Phillips Memorial Cloister was recognised as a Grade II Listed Building in 1991. Despite this acknowledgment, the Cloister suffered from vandalism and deterioration, both of which were noted periodically in letters to Waverley Borough Council and local newspapers. In an attempt to revive the site, and in order to commemorate the 150th anniversary of the birth of Gertrude Jekyll, the Surrey Gardens Trust restored the planting in and around the Cloister and reinstated the pond in 1993. This was achieved through donations of plants and assistance from other organisations and individuals, demonstrating continued interest in the Park's heritage. At the same time, minor repairs were made to the building, lighting was added, and two interpretation boards were erected on the site with a further five mounted on the walls of the Cloister.

SITA Environmental Trust awarded Waverley Borough Council a grant of £5,000 in 2000, in order to increase the visual appeal and conservation value of the meadow area and to encourage its wider use as an educational resource. As part of the project, Godalming Scouts, Moss Lane School and Godalming Town Council were involved in wildflower plug-planting.

In furtherance of the aims of the SITA project, and as a solution to path flooding, a scrape was created in 2003, near the Council Offices, making use of natural seepage. This allowed running water, which previously cascaded unchecked over the Riverside Walk, to be collected and to provide a habitat for amphibians and other pond life. A boardwalk around the outside was added to allow pond dipping and better access for litter removal.

A resurgence of interest in the Park in recent years has been sparked through the introduction of a calendar of summer events. Central to this was the establishment of a formal bandstand in 2005, echoing a previous feature of the Park. The earlier structure was created in 1914, near to the Bowling Green, using soil excavated from the site strengthened by concrete (formed using shuttering previously used to shape the raised pond); this provided a raised platform for musical recitals. In 2003, the GO Godalming Association held an open air concert on the existing plinth in the park, which proved so popular that five further concerts were organised in 2004. However, the plinth proved too small and was considered dangerous for musicians. In 2005, the two local Godalming Rotary clubs worked with GO Godalming to provide a new bandstand, based on a similar structure in Cobh (the last port of call for the

Titanic). The enlarged plinth, faced with Bargate stone, is paved with York stone in the form of a Rotary Wheel, and is fitted with railings around the perimeter. A roof was added in June 2009. (Fig. 5) The whole bandstand is now the property of Godalming Town Council, although responsibility for the audience area remains with Waverley Borough Council!

Anniversaries

Godalming Town Council has arranged services of remembrance in the Park for the major anniversaries of the sinking of RMS *Titanic*, including the 75th anniversary held in 1987, followed by the 85th when more than 200 people gathered (including members of the British and Irish Titanic Societies); and more recently in 2002 and 2007, on the occasions of the 90th and 95th anniversaries. The services have included prayers, hymns and anthems performed by Godalming Band and a combined choir from Godalming and Farncombe churches. The 100th anniversary has been commemorated in a similar manner, recalling the original opening ceremony.

Recent history

The 100th Anniversary of the sinking of the Titanic on 15th April 2012 provided an opportunity to make sure that the various heritage features of the Park were restored, conserved and enhanced for the benefit of future generations. Maintaining an historic feature like the Park is expensive, and Waverley Borough Council was very fortunate to be awarded, in July 2011, a Parks for People grant by the Heritage Lottery Fund and the Big Lottery to cover a five year programme of restoration and enhancement. The success of this funding application was in no small way due to the huge support given voluntarily by members of the Steering Group, who offered support and guidance throughout the application process. Of particular note, representatives from the Surrey Gardens Trust produced a new, costed planting scheme, based on Jekyll principles, for the gardens in and around the Cloister. It is hoped that voluntary support will continue to be given through the Steering Group, acting in the future as a consultative group, and also by encouraging individuals to become actively engaged in practical volunteer work on the ground. A further intended development is the ultimate creation of a Friends Group to support work in the Park.

FIG. 5
THE BANDSTAND IN THE PHILLIPS MEMORIAL PARK, GODALMING STAYCATION
(PHOTOGRAPH: NICK BAXTER)

Restoration of the Cloister has included the repair of the oak pergola, along with underpinning of walls and uprights in the main part of the Cloister. Roof tiles have been

Fig. 6
The Cloister during restoration (Photograph: Sarah Sullivan)

replaced and inappropriate materials, for example cement mortar used for pointing during less sympathetic periods of maintenance, have been replaced. (Fig. 6) Finally, low energy architectural lighting has been installed to highlight the beauty of the building on special occasions. Other intended work across the Park includes improvements to physical access, plus enhanced interpretation of the site and biodiversity improvements.

Fig. 7
Nursery children building a snowman, December 2010 (Photograph: Lizzie Noble)

As part of the grant application, a 10 Year Maintenance and Management Plan has been produced to ensure that work in the Park is implemented in a structured manner in the future, rather than the sometime ad hoc approach adopted in the past. This will build on the Masterplan, produced specifically for the five year programme of grant-aided enhancement works, and will provide continuity in management, for example by guiding the choice of trees for the site. It is intended that the Plan will be reviewed and updated regularly, and that quinquennial reviews of the Cloister will also be completed to ensure that the building receives appropriate regular maintenance.

The Park is here to be enjoyed, and it is intended that use of the Park will increase. Already the Park provides the venue for Godalming Town Day and Farncombe Parish Day, and it is anticipated that the hugely popular summer concerts (organised by GO Godalming and Godalming Town Council) and the Staycation events (organised by Godalming Town Council) will continue to bring more people into the Park in the summer months. More recently, the Park has been used by Godalming Scouts as the venue for their St George's Day commemoration (2011) and for a modern-day Christmas Nativity Play, performed by Godalming College (2010). It is also hoped to develop opportunities for other events, such as art exhibitions and musical performances, to take place across the Park and inside the Cloister, as was originally intended. (Fig. 7)

Ultimately, the success of the current project will be proven, if the Phillips Memorial Park continues, as envisaged originally by Gertrude Jekyll and Hugh Thackeray Turner, to be:

Something beautifying to Godalming and also both useful and enjoyable

The restoration of the Phillips Memorial Park is supported by the National Lottery through the Heritage Lottery Fund and the Big Lottery Fund.

Fig. 8
Phillips Memorial Cloister with original bench (John Young collection)

A TIME LINE OF MAJOR EVENTS WITHIN THE PHILLIPS MEMORIAL PARK

Date	Event
Pre-1913	Glebe land/Open grazing land
1912	A committee established to oversee the development of the park
1913	3 acres purchased for the purpose of the Phillips Memorial Garden
1914	The Phillips Memorial Cloister unveiled (15 April)
1920	Town War Memorial erected (designed by Powys)
1924	Godalming and Farncombe Bowls club established in the Park
1926	Sanitary conveniences installed (later removed)
1933	Refectory Manor Animal Pound given to the people of Godalming (the Phillips Memorial Gardens)
1952	More land purchased (The Burys Youth Field)
1956	Nine weeping willows removed
1959	More land purchased (The Burys)
1964	Four poplars between the Cloister and road removed
1965	Tiles on the north and west side of the Cloister replaced. South facing wall of the Cloister replaced with a pergola. Replanting carried out.
	Riverside Walk constructed
1966	3 parts of the site brought together and renamed 'The Phillips Memorial Park'
1974	Re-organisation of local government and the establishment of Waverley Borough Council and Godalming Town Council.
	Establishment of the Godalming Conservation Area
1984 - October	Designated Conservation Area extended
1986	The pond was filled in by students from Charterhouse

1987	75th Anniversary of the sinking of RMS Titanic
1989	Playground for children installed
	Designated Conservation Area extended (December)
1991	Phillips Memorial Cloister is listed Grade II
1993	Surrey Gardens Trust restore the Jekyll planting and reinstate the pond to commemorate the 150th anniversary of the birth of Jekyll
	Minor repairs to the Cloister
	Installation of interpretation boards within the Park
1997	85th Anniversary Event
2000	SITA Environmental Trust grant for meadow area improvements
2002	90th Anniversary
2003	Construction of a scrape and boardwalk
2005	Bandstand reinstated by Go Godalming
2009 March	Steering Group established
2010	Awarded Parks for People stage one grant
	Development Stage Team established in October to prepare a Stage 2 application for funding.
2011 July	HLF and Big Lottery awarded a Stage 2 grant of £335,000 towards a 5 year programme of work intended to restore and enhance the Phillips Memorial Park, for the enjoyment of future generations

BIBLIOGRAPHY

The Arts and Crafts Movement in Surrey (eds) *Nature and Tradition, Arts and Crafts Architecture and Gardens in and around Guildford*. The Arts and Crafts Movement in Surrey and Guildford Borough Council, 1993, second edition 2002

Brown, Jane, *Gardens of a Golden Afternoon*, Viking, 1982

Brown, Jane *The Story of a Partnership: Edwin Lutyens and Gertrude* Jekyll Penguin Books,1985

Budgen, Christopher, *West Surrey Architecture 1840-2000.* Heritage of Waverley Ltd., 2002

Jekyll, Francis, *Gertrude Jekyll: A Memoir* Jonathan Cape, London, 1934

Jekyll, G and Weaver, Lawrence *Gardens for Small Country Houses*. Fifth Edition. Country Life, London, 1924

McCluskie, Tom; Sharpe, Michael & Marriott, Leo, *Titanic & Her Sisters Olympic & Britannic* PRC Publishing Ltd, London, 1998

Nairn, Iain & Pevsner, Nikolas, *The Buildings of England – Surrey*. Penguin Books 1999

Tankard, Judith B and Wood M.A., 1996 *Gertrude Jekyll at Munstead Wood*. Stroud, Sutton Publishing Ltd., 1996

Ticehurst, Brian, *Titanic Memorials World-Wide 1996*. B & J Printers, 1996

Tooley, Michael, 'Gertrude Jekyll (1843-1932), artist and garden designer'. *Oxford Dictionary of National Biography*. Oxford, 2004

Tooley, Michael and Arnander, Primrose (eds) *Gertrude Jekyll: essays on the life of a working amateur.* Michaelmas Books. Witton-le-Wear, Co.Durham, 1995

Pamphlet

Jekyll, Gertrude, *Public Parks and Gardens*, Civic Arts Association, 1918

Auction Catalogues

April 14th 1992: Christies Maritime Auction – Telegraphic messages

May 15th 1997: Christies Maritime Auction – Elsie Phillips Postcard Album Lot 20, realized £2200

Periodicals

January 12th 1912, Country Life *Westbrook, Godalming the Residence of Thackeray Turner* by Gertrude Jekyll

April 19th 1912, Guildford and Godalming Weekly Press – *Duty Remembered*

April 20th 1912, The Surrey Advertiser and County Times (Third Edition) - *Surrey Victims*

April 20th 1912, The Daily Graphic. *Titanic – in Memoriam* Number

April 27th 1912, The Surrey Advertiser and County Times (Third Edition)

April 17th 1914, The Surrey Weekly Press – *Memorial to a Man*

July 24th 1918, Country Life. *In the Garden. Mr Thackeray Turner's Garden at Westbrook*

June 14th 1924, Gardening Illustrated, *The Phillips Memorial at Godalming* by Gertrude Jekyll

GAZETEER

Local places to visit associated with Jack Phillips and the Arts and Crafts Movement in Surrey
Check current opening times on websites

GODALMING MUSEUM

109A High Street, Godalming, Surrey GU7 1AQ
www.waverley.gov.uk/godalmingmuseum   Tel: 01483 426 510
Permanent Gertrude Jekyll display and an extensive Jekyll archive in local studies library.
Thackeray Turner ceramics.

GUILDFORD MUSEUM

Castle Arch, Guildford, Surrey. GU1 3SX
www.guildford.gov.uk/museum   Tel: 01483 444 751
Holds Jekyll's Old West Surrey collection, partly displayed. Thackeray Turner ceramics in store.

WATTS GALLERY

Down Lane, Compton, Surrey GU3 1DQ
www.wattsgallery.org.uk   Tel: 01483 810 235
Memorial collection of George Frederick Watts with material on Mary Seton Watts.

VANN

Hambledon, Godalming, Surrey GU8 4EF
www.vanngarden.co.uk   Tel: 01428 683 413
Jekyll water garden.

ST JOHN THE BAPTIST CHURCH

Busbridge, Godalming Surrey GU7 1XA
www.bhcgodalming.org   Tel: 01483 421 267
Gilbert Scott Church, Burne-Jones windows and Lutyens Chancel screen. Jekyll's Parish Church and grave.

ST JOHN THE EVANGELIST CHURCH

St. Johns Street, Farncombe, Godalming Surrey GU7 3EH
www.farncombe.org.uk   Tel: 01483 426 353
Gilbert Scott Church. Phillips's Parish Church with brass memorial tablet in north isle.
Phillips sang in the Church choir and attended adjacent school (now Farncombe Day Centre)

NIGHTINGALE CEMETERY

Nightingale Rd, Farncombe, Godalming Surrey
Tel: 01483 523 575 (Clerk to the Godalming Joint Burial Committee)
Iceberg tombstone to John George (Jack) Phillips along with the names of his father, mother and twin sisters on the stone perimeter slabs that surround the berg.

ACKNOWLEDGEMENTS

The Society for the Arts and Crafts Movement in Surrey acknowledges with great pleasure the enthusiasm and generous assistance received from both individuals and organisations during the completion of this project. In particular, thanks are due to the contributing authors for their essays and for the permission to use illustrative material, either from their own archive or collection, such as that of John Young, or created by their own hand, as seen in Sarah Sullivan's evocative drawings. Alison Pattinson and the staff at Godalming Museum have been most supportive, as have Mary Alexander at Guildford Museum and Mark Bills and the Watts Gallery staff. Gail Naughton and Ann Laver have been invaluable in researching the Phillips Family Tree, Justin Voisin has permitted use of images from a Westbrook photograph album and John Fairbanks has given kind consent to include his drawing of Red House.

Help with research was also kindly given by History Tutors – Nick Swan at Aldro School and Mrs Miller at Priors Field School.

Credits to Organisations: The Society for the Arts and Crafts Movement in Surrey Acknowledges:

Permission granted to reproduce copyright material in this book. Every effort has been made to trace copyright holders and obtain their permission for the use of copyright material. The Society and the authors apologise for any errors and omissions and would be grateful to be notified of any corrections that should be incorporated in future reprints or editions of this book.

Jekyll Drawings:	Godalming Museum Local Studies Library (Courtesy of Gertrude Jekyll Collection (1955-1), Environmental Design Archives, University of California, Berkeley)
Jekyll Plant Notebooks:	Godalming Museum Local Studies Library
Newspapers:	Surrey Advertiser Historical Archives at Surrey History Centre, St Johns, Woking

Russell Morris, Sarah Sullivan, Nick Baxter and John Young at the Phillips Memorial Cloister (Photograph: Jayne Fincher)

THE AUTHORS

Russell Morris:
An historic buildings officer within Surrey and West London for some twenty years. Besides an interest in Arts and Crafts architecture, he is a compulsive collector of antiquarian and early 20th century books on building construction.

John Young:
A building surveyor. He lectures and writes about the Phillips Memorial and volunteers at the Godalming Museum. John is a committee member of the Surrey Postcard Society and has a vast collection of material relating to Jack Phillips, Titanic and images of Godalming and the surrounding area.

Amanda Le Boutillier:
Contributing Associate Member of the Titanic History Society, Jersey since 1995.

Michael Tooley:
Lectures throughout the UK, Ireland and North America. He has written more than a hundred articles and chapters in books on aspects of the environment, gardens, garden history and contributed the entry on Gertrude Jekyll to the *New Oxford Dictionary of National Biography*.

Sarah Sullivan:
Heritage and conservation specialist for ADAM Architecture with a particular interest in timber framed buildings. Upark (National Trust) is amongst the many building restoration projects she has worked on. Sarah is currently researching the life and work of the architect Charles Harrison Townsend (1851-1928).

Robin Stannard:
A historic buildings surveyor working for ADAM Architecture. Robin began his conservation career with SPAB architect David Nye and then worked for English Heritage. With a special interest in the Arts and Crafts Movement, he has spent several years researching the life and work of Hugh Thackeray Turner.

Nick Baxter:
A Chartered Biologist and a founder member of the Institute of Ecology and Environmental Management. He has worked in countryside management for over 30 years, and has considerable experience of working with, and establishing, volunteer groups. He has been the lead on three successful HLF bids, the most recent being to restore the Phillips Memorial Park.

Lizzie Noble:
With an MSc in conservation biology, and brought up in Godalming she has, together with Nick, formed the team that submitted the successful Parks for People funding bid. More recently, she has been credited with the rediscovery of the Santa Maria Toro *Santamartamys rufodorsalis*, an arboreal mammal living in the Colombian cloud forest, which had not been seen since 1898.